# LOTUS SEVEN

# LOTUS SEVEN

Andrew Morland

First published in Great Britain in 1994
by Osprey, an imprint of Reed Consumer
Books Limited, Michelin House,
81 Fulham Road, London SW3 6RB and
Auckland, Melbourne, Singapore and Toronto.

ISBN 1 85532 490 3
Project Editor Shaun Barrington
Editor Simon McAuslane
Page design Paul Kime/Ward Peacock
Partnership

Printed and bound in Hong Kong
Produced by Mandarin Offset

# ACKNOWLEDGEMENTS

A very special thank you to the owners and enthusiasts of these
entertaining cars for their time, cooperation and help in
photographing them in all weathers.

Thanks in particular to Jez Coates of Caterham Cars, Bruce Robinson
of Arch Motor and Manufacturing Co., Joop Donkervoort of
Donkervoort Automobiel, Chris Smith of Westfield Sports Cars Ltd,
Vincent Haydon of Supersprint Cars, Peter Henshaw and, lastly but
not least, the enthusiastic members of the Lotus Seven Owners' Club.

**Title page**
*Just to keep you on your toes...*
*No, it's not a Seven. See page 122*

For a catalogue of all books published by Osprey Automotive
please write to:

**The Marketing Department, Reed Consumer Books,
1st Floor, Michelin House, 81 Fulham Road, London SW3 6RB**

# Introduction

Anthony Colin Bruce Chapman, at the age of nineteen, rebuilt a 1930 Austin Seven Saloon into a special and called it a 'Lotus'. The Mk 1 and Mk 2 Lotus cars were built for trials. The Mk 3, built with his friends Michael and Nigel Allen, was a 750cc Austin engined car for the 750 Formula racing popular in Britain in the 1950s. The Mk 3 was built in 1951 in the Allens' garage at Vallance Road, Wood Green in North London. It won its first race at Castle Combe and went on to win many races in the hands of Colin, the Allens, and Hazel Williams, later to become Chapman's wife.

The Mk 3 had a very powerful 750 Austin engine 'de-Siamesed' by Chapman. The original Austin 7 chassis was strengthened by triangulated tubular construction, which made it rigid and light.

The Mk 6 had evolved from the Mk 3 as well as from Colin's experience of 750 racing. The 750 formula was full of talented improvisers and engineers, such as Eric Broadley, Jack French, Len Terry and Arthur Mallock. Colin used the Ballamy swing-axle front suspension for the Mk 6 as it worked on Ballamy's Ford Specials.

*Production line of Super Sevens at the Caterham factory in Dartford, Kent*

# Contents

**Left**
*Lotus Mk 6 with owner and restorer James Atkinson at the wheel*

# Lotus 6

The Lotus 6 was the first production car from the Lotus Engineering Company based at Hornsey, North London. It was sold from 1953 to 1955 in 'do-it-yourself' kit form to take mainly Ford 10 parts. Austin 7 and Ford 1172cc specials were popular in this period, but Colin Chapman offered a much more advanced and remarkably light car. Colin's multi-tubular space frame chassis was designed for low weight and great strength; rigidity and stress was calculated mathematically. The result was the Mk 6 chassis weighing only 55 lbs. The lower main chassis tubes are largish one-and-seven-eighths diameter 18 gauge with upper sections in one inch rounded and square tubes; extra rigidity is provided by riveting aluminium panels to the floor and frame. Even with these

**Above**
*Coventry Climax 1460cc with twin Weber carburettors fitted to Tom Candlish's Lotus 6. In the up-to-1500 class of racing the MG TF 1497 was a much more popular and cheaper engine for the Six*

**Right**
*Lotus Mk 6 with owner Tom Candlish in the paddock at Prescott Hill Climb. This 1955 model is fitted with the rare four cylinder Coventry Climax 1460cc engine*

stressed panels fitted it still weighs only 90lbs, which is extraordinarily light by any standards.

The Lotus 6 chassis was built by Progress Chassis Company, Edmonton, North London, run by John Teychenne, who used to work for Colin Chapman. Working with John were two skilled chassis builders, Dave Kelsey and Mike Coltman.

At Progress the chassis numbers were stamped for Colin Chapman on the front suspension bracket mounted on the bottom down tube. However, not all the numbers are easy to read as they were hand punched. The story goes that for about half the production, the numbers were stamped on the completed chassis. As the chassis was light and springy the hammer would spring back and the numbers were difficult to identify. Eventually it was decided to stamp the bracket on the work bench to give a good clear result.

Williams and Pritchard, also based in Edmonton, built the beautiful aluminium bodies for the Mk 6. These skilful body builders worked closely with Lotus Engineering on the initial design of the Mk 6. The bodies are quite complicated and not as simple as they appear.

The 'touring' back, which includes the hood frame for the road car, is a very clever and skilled piece of skinning. Those re-skinning their 6's today cannot believe the work involved in this area of the body.

One interesting variation on Mk 6's that makes it easy to distinguish an early car from a later one is the length of the bonnet and scuttle top; the later cars have a shorter bonnet and a larger skuttle. This was changed to make it easier to fit the single aero screen or perspex on the scuttle for those racing.

Colin Chapman's clever use of cheap, easily available Ford components kept the overall cost of a completed Mk 6 down to an affordable price. Customers could purchase their own Ford parts and modify them, or get Lotus to do it for them.

The E93A Ford front axle was cut in half and two plates welded on the ends into which bushes were fitted. The Ford 'V'-shaped tie rod was cut in half and shaped to form radius arms. This form of swing-axle front suspension was also used by Ballamy on his Ford specials. The Mk 6 uses the Ford steering box and column, which does tend to spoil the driving pleasure due to its tightness, play and lack of feel.

It seems fairly certain that Colin Chapman would have thought about rack and pinion steering, which was already common in the 1950s. However the use of the swing axle front suspension makes the fitting of rack and pinion difficult, without causing bump steer. The later Lotus Mk 8 racer with swing axle suspension used a combination of rack and pinion and Burman steering box. Some owners did manage to fit rack and pinion steering, such as the type fitted to Chris Smith's Mk 6.

**Right**
*Fred and Lee Fairman working on Lotus 6 with touring body and pretty enclosed mudguards. Fred and Lee are expert chassis and body constructors for Lotus 6 and early Sevens*

**Below right**
*James Atkinson's Lotus Mk 6. Balanced 1466cc Wolseley/MG TF engine, Laystal head, MG TC gearbox, hydraulic brakes, chassis no. 62*

The Mk 6 uses the Ford 10 back axle located by a Panhard rod to the chassis. The Ford Torque tube was shortened but hub wheels and gearbox were standard.

The Ford 10 back axle was always weak – the gears would break and half shafts snap. Today it is possible to improve the axle with kits for the rear hub carrier. Most Mk 6's have the special 4.7 ratio differential though Ballamy, the Ford Special man, could supply 4.4 and 4.1 ratios. Many Mk 6's have the Ballamy hydraulic brake conversion, which makes 100 per cent improvement to the stopping distances. Ballamy also supplied the special Ford 15-inch wheel for the Mk 6, though some owners ran the original 17-inch Ford wheels on the front axle.

How many Lotus 6's were built? Despite the efforts of the Lotus 6 registrar, Charles Helps, the definitive number is not known as no specific records were kept. The number has always been suggested at

**Above**
*Lotus 6 built in 1955 with the 1460cc Coventry Climax engine at Prescott Hill Climb, on Ettore's Bend with Tom Candlish at the controls*

**Right**
*James Atkinson's beautifully restored Mk 6 was owned by the author in the early seventies and first registered on the road, 21 May 1955*

just over 100, or including the Mk 8, 9 and 10, maybe 150. In 1954 Colin Chapman decided to reserve the Mk 7 number for the improved version of the Mk 6, which appeared in 1957.

The checking of chassis numbers is not helped by Mk 8 racing cars using the 6 chassis, sometimes with an added prefix number visible. Only seven Mk 8's were built and all are accounted for except one.

The Mk 8 was basically a Mk 6 chassis with extra tubes to support the aerodynamic aluminium body designed by Frank Costin, the Aerodynamist who worked for the De Havilland Aircraft company. This body increased the Mk 6's top speed by about 20mph.

The Mk 9 racing car used a '6' type chassis but the dimensions of the tubes and lengths were altered to save weight and still support the stream-lined body. The Mk 9 chassis retained the wheelbase of the Mk 6 at 87.5 inches. The 1955 Motor Show Mk 9 had the chassis no.130 and was raced by Peter Lumsden.

Deducting numbers of Lotus racing cars built would give the Mk 6

**Left**
*Lotus 6 Ford engine, built in 1954, racing round Ettore's bend at Prescott, with owner Philip Rushforth at the controls*

**Above**
*The Ford Consul 1498cc engine, first time up hill after restoration, seems to be coping pretty well*

*Lotus 6, 1954, chassis no. 38. Has original seat layout but a Williams seat belt for racing*

production run a figure nearer 115. In addition to these numbers are some six chassis that were registered in the 1950s which Progress Chassis Company have disclaimed building. Did Colin Chapman employ someone else or was he unaware that someone was building the odd extra chassis? The Mk 10 chassis, despite similarities to the Mk 6 chassis with its 87.5-inch wheel-base, is very different with the enlarged engine bay for the 2-litre Bristol or Connaught engine.

## Racing Lotus Six

Sinclair Sweeny's first Mk 6 had the solid-beam Ford Popular front axle, with high ground clearance and had great success in trials and rallies.

All the other Mk 6's had the Ford front axle cut and the centre plates

*Ford Consul engined Lotus 6. The dashboard has a faired cowl not used on the Sevens and there is a more recent Lotus gear knob. Note period Williams and Pritchard badge, makers of the aluminium bodies*

welded and bushed to pivot. The V-shape of the original tie rod was rebuilt into two radius arms. This independent swing axle front suspension, when set up correctly, could give a vertical wheel angle. Though photographs from the 1950s of Mk 6's racing show many front wheels at a weird angle.

There were many successful racers in their Lotus Mk 6's of the period. In the 1172cc Ford-based cars Ken Laverton and John Lawry were tops, while Mike Antony was winning in the MIGTF-engined Six.

Fred Hill of Empire Garages had a most unusual Six with a 1933 supercharged 746cc MG J4 engine. In 1953 he was winning cups and the car could be seen at his garage in Finsbury Park. Bill Perkins used a BMW 2 litre with success. John Harris had a beautifully prepared 1100cc Coventry Climax engined car with de Dion rear axle and large turbo finned alloy brakes – he took many class wins. Not forgetting Colin

Chapman and Nigel Allen who raced 6's reg nos. 1611 H and XMI 6.

However, the most successful driver of the Lotus 6 was Peter Gammon. A draper by profession, Peter was a very talented and professional racing driver; he raced a very quick MG TC special. Peter actually spoilt the Lotus 6's first racing appearance at the MG Car Club meeting at Silverstone by beating Colin Chapman, taking first place with his MG. However, Peter Gammon was so impressed by the Lotus 6 he immediately ordered one from Colin Chapman. Peter Gammon's beautifully prepared 6, reg. no. UPE 9 with an MG TF 1497cc engine, won 17 events in the 1500cc class in 1954, against stiff opposition including Cooper, Tojeriro and Connaught racing cars. Peter also won the Performance Car Trophy in 1954.

One wonders how many Lotus 6's were not used in some form of competition.

The Mk 6 that I owned for six years and in which I travelled over 50,000 miles in the early 1970s was raced by Sir John Whitmore. The Registration number was SPW 990, chassis no. 62. Today it still has its fully balanced 1466cc Wolseley 4/44 engine with XPEG MG TF pistons, Laystal aluminium head with 36mm exhaust valves, twin 1.5 inch SU carbs and Derrington four branch exhaust, giving around 85bhp. Sir John Whitmore also raced another Mk 6, reg EDC 844, which has had a full chassis and body rebuild to a very high standard by the Mk 6 specialists Fred and Lee Fairman.

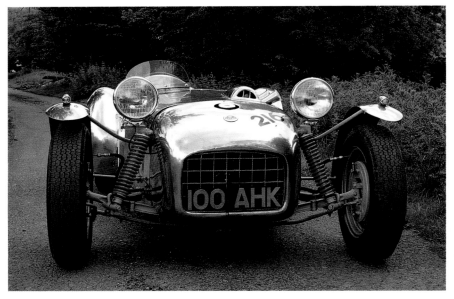

**Top**
Lotus 6 chassis being held by Ace Lotus 6/7 restorer Mike Brotherwood. This demonstrates how light the chassis construction is

**Above**
Lotus 6 chassis no. 38 with Laystall tuned Ford Consul engine, beautifully restored by Philip Rushforth

**Right**
Lotus 6 Ford Consul engine chassis no. 38, with owner Philip Rushforth of Abberley, Worcestershire

Chris Smith racing his famous Lotus Mk 6, built in 1954, at Mallory Park. This is the ultimate Six, with an MG XPAG engine with Laystall Head, de Dion rear with leading parallel links, enclosed rear wings and rack and pinion steering. Chris Smith is the owner of Westfield Sports Cars

*The Lotus 6 became the standard club car of the 1950s and made the name famous worldwide. The Seven took a similar role from the 1960s onwards. Colin Chapman often displayed a healthy disregard for accepted rules of design and often said that: "the trouble with experts is they know what can't be done.". In his book* Colin Chapman Lotus Engineering *(Osprey, 1993), Hugh Haskell says: "This is of course another way of expressing the better known phrase 'an inventor is someone who doesn't know that what he is trying to do is impossible'."*

# Series One
# The Bread-and-Butter car

Her Majesty's Inland Revenue was responsible for the Lotus Seven. Well, maybe that's a bit of an exaggeration – Colin Chapman did have a hand in it as well. He could recognise a loophole better than most, particularly one which would enable him to produce a little sports car that not only outperformed the mass-produced offerings, but undercut them as well.

The loophole in question was Purchase Tax, payable on all new cars but not car components. And to someone like Chapman, never coy concerning legal circumvention of the rules, the answer was obvious; sell the car as an outsize Airfix kit, for the owner to complete its final assembly. Thus tax would be avoided and the natural enthusiasm of a typical Lotus owner harnessed in one fell swoop!

It was a different approach to the Lotus 6, which sold as a basic body/chassis kit, depending far more on the owner's ingenuity to find suitable engine, gearbox and suspension wherever he cared to look. But there were of course many other reasons for the Seven's existence, especially Chapman's ambitions for his fledgling company. By 1956, it was well established as a maker of very light and fast (if a little fragile) racing

**Left**
*Lotus Seven Series 1 chassis. A newly rebuilt chassis by Mike Brotherwood*

**Right and overleaf**
*A BMC 'A' series-engined Lotus Seven Series 1 restored by Brian Anthill of Corscombe, Dorset, who specialises in early Lotus Seven restoration*

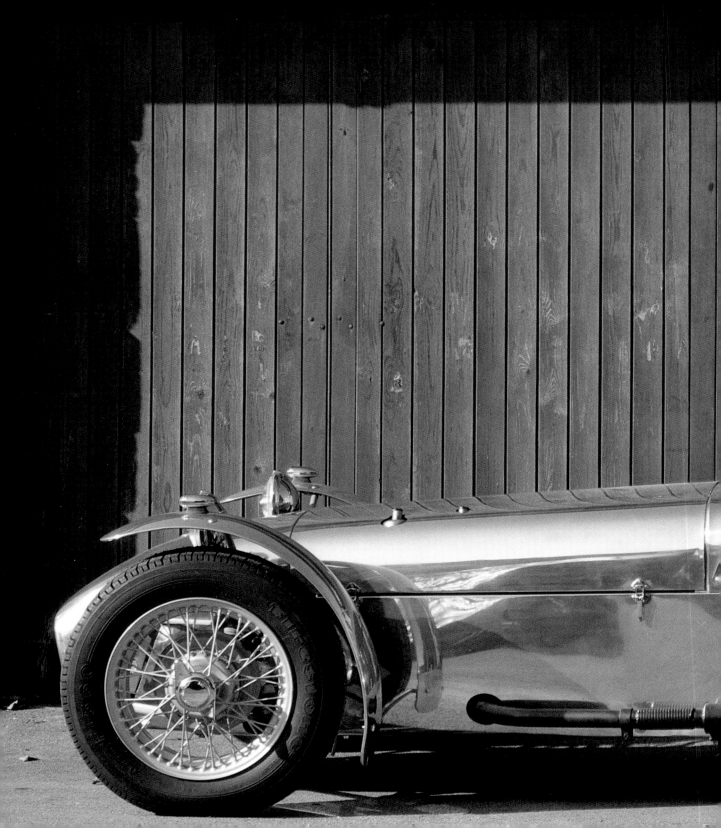

cars. But the young designer wanted to go on to other things.

Specifically, he wanted to produce a pure 'road' car which would incorporate all the accumulated experience and expertise derived from racing. It would be innovative as well as highly efficient – and it would eventually come to fruition in the Elite. But development of the cutting edge of technology (especially if you're selling to the public rather than racing drivers) took time and money, even in the mid-fifties. To finance such a project, Lotus needed a bread-and-butter car, something cheap and simple which would sell all year round and bring in a steady income.

And that's just what the Seven did. It was closely based on its predecessor (which actually ceased production in 1955), but was lighter, simpler and easier to build. The basis was still a clever spaceframe, by now a Chapman trademark, which set it apart from all the other kit cars (and indeed from all the MGs, Triumphs and Morgans). A web of mostly

*Lotus Seven Series 1, 1959, owned by Brian Anthill. This rear shot shows the pretty, hand-beaten alloy wings, the spare wheel and its carrier with number-plate mountings*

**Above**
*Lotus Seven Series 1, circa 1959. Note the Cycle wings with side lights and the original Lucas spot lights with single filament bulbs used for headlamps. Larger than standard SU. H4 carbs protrude through the bonnet*

**Overleaf**
*The same Lotus Seven Series I built in 1959. Early advertisements for the kits promised that the vehicle could be built in just 60 hours*

one or 3/4-inch tubing, the spaceframe was simplified from the Six. More body panels (and there weren't many on a Seven) were riveted to it thus stiffening up the frame and enabling Lotus to remove some of the tubes.

This wasn't just to save money – from the start, the Seven was to embody a fundamental Lotus (i.e. Chapman) truth. Speed came not from lots of power, but lack of weight. Colin Chapman's work at British Aluminium, stressing bridges, was put to good use in the Seven's spaceframe. It was properly triangulated, with all the different loadings calculated (by pencil, paper and slide rule in those pre-CAD days). The result was very light, weighing slightly less than half a hundredweight and it was very strong – well able to provide the excellent handling which Lotus customers expected.

All of this was actually little advanced on the Lotus 6, but things were different in the suspension. Gone was the 6's sawn in half Ford beam

**Left**

*Lotus Seven Series 1, 1959, with optional 15-inch wire wheels shod with 155 x 15-inch radials. The original fit would have been 145 x 15 radial or 520 x 15-inch crossply tyres. Extra orange indicators were often fitted by owners but many used side lights in front wings and red rear lights as indicators*

**Right**

*Lotus Seven Series 1, 1959. BMC based car owned by Brian Anthill. It is fitted with a much later 'A' series 1275cc 'Spridget' engine from 1965. This gives the Series 1 much more power especially with the oversize 1.5 inch twin SU H4 carburettors. Besides the oversized carbs and modern fuel lines the engine is very similar to the original 948cc. The A-series BMC Lotus Seven was always quicker than the Ford 100E engined car which only had a three speed gearbox to the A-series four speed*

axle. Instead, a proper independent system graced the front, derived from the Lotus 12 F2 car. Double wishbones were used, with steeply cambered coil spring/damper units which have become the trademark of every Seven since. There was also an anti-roll bar, which doubled as the forward member of each upper wishbone. The result was cheap, effective and easy to make.

Things weren't quite so clever at the back, which supported a Standard 10 live axle. It. was well located (by two trailing arms and one diagonal) and Lotus was able to offer an impressive list of ratios to fit it. But the cornering powers of any Lotus Seven are rather more than those of a Standard 10, putting untold stresses on the axle and diff casing – the resultant breakages made this component the Seven's weakest link, though strangely it wasn't replaced by a stronger one until 1958. Steering started off with a Burman worm-and-nut set up, but soon after production started, went over to the Elite's rack-and-pinion: much nicer.

It was almost as if, having given the Seven an advanced space frame, Lotus had to use a cheap engine to save money. Ford's infamous 1172cc side-valve unit was a sensible choice though. It was the 'standard' fitment to the Lotus 6. Already familiar to a generation of specials builders, it had plenty of scope for tuning. And anyway, the whole Seven concept was that you didn't need lots of expensive power in order to have fun. In standard tune, the 1172 gave 40bhp and came with a wide ratio three-speed gearbox. It was in this form, in October 1957, that the Seven Series One was presented to the public at the Motor Show. The Elite

made its first appearance there too, but it was still some way from production. The price was just £525 in kit form, but if you insisted Lotus would build one for you. Including labour and Purchase Tax, this came to £1,036/7d. Not surprisingly, most Sevens left the works as a box of bits.

But to get that keen price, the first Seven's specification was, shall we say, minimalist. A full-width windscreen was standard, but that was about all. Hood and tonneau cost extra, as did windscreen wipers, while Lotus

**Above**

*Lotus Seven Series of 1959. Good view with nose cone and bonnet removed and showing mounting of bottom wishbone of front suspension and antiroll bar used as a top suspension arm; a great design by Chapman for simplicity and light weight. However as the Seven became faster a proper top suspension wishbone was needed for quick and safe handling and a better ride. The old Series 1 set-up lasted until 1985 when Caterham offered an extra half wishbone to provide negative camber on the front wheels for their de Dion cars*

**Right**

*Lotus Seven Series 1, 1959. Close up showing front suspension which despite being inferior to a double wishbone system was lighter; it was well up to the power and tyre grip of the period. The antiroll bar that is bushed into the top link restricted suspension movement. Damping is controlled by the coil spring shock absorber. This front suspension came from the 1956 F2 Lotus racing car which Graham Hill and Cliff Allison raced successfully in 1957*

wouldn't supply a heater however nicely one asked! Rev counters, fuel gauges and side screens were likewise dismissed as fripperies. Like the man who asked the price of a Rolls-Royce, anyone wanting a waterproof Lotus Seven would be advised to go elsewhere.

The first road tests, in their discreet 1950s style, were never so vulgar as to offer any real criticism. (For example: 'The exhaust note at high engine revolutions is noticeable'.) In fact, *Autocar* concluded its December 1957 test with, 'For the enthusiast with a desire for racing, the Lotus Seven is a safe and sensible vehicle.' The test car had the optional tuned side-valve (twin SUs and a four-branch exhaust manifold) and close ratios in its three speed box. The result was a surprisingly spirited performance – first would reach 85mph, and second would stretch right up to 70. 0–60 in 17.8 seconds sounds rather less than impressive these days, but it was good enough for 1957, '...one soon forgets minor discomforts in the exhilaration of its performance, and the manner of its achievement'. Top speed was slightly disappointing, at 81mph, pointing to a perennial Seven characteristic: aerodynamically, they have always been closer to the brick wall than the teardrop school. But none of this really mattered – then or now; Sevens are not about 100 mph-plus speeds, but about how it feels on the way up to eighty.

However, some people did want more performance, and an 'Export' model soon appeared, with a mildly tuned, 48bhp version of the 1172. It also had the close ratios, a spare tyre (!) and optional wire wheels. But it was over a year before serious performance became available. In December 1958, the first Super Seven appeared, using the familiar Coventry Climax FWA as motive power. It was in a different league to the old Ford side-valve – all alloy, overhead cam, racing heritage...the little Climax had actually started life as a fire pump, but it still gave 75 bhp from only 1098cc. This naturally gave the lightweight Seven startling performance – 0-50 In less than nine seconds, and a highly theoretical 130mph on the right rear axle ratio. It naturally cost a bit more than the basic version – £700 as a kit – but had a number of extras. Four speed gearbox, wire wheels, rev counter and a leather ccvered steering wheel which replaced a white, plastic two-spoke affair. Much more suitable.

Yet another engine option appeared the following year. The Seven A used BMC's 948cc A-series engine of 37bhp. Offering about the same power as the side-valve, its main benefit in the UK was to give impecunious Seven buyers the choice of a more modern engine. The real

*Lotus Seven Series I, 1959. Note the fuse box in front of the passenger and no rev counter. It was not offered in the kit and owners of series I's had to fit their own. Steering wheel is the type common on the Series 2*

who hankered after more powerful and much faster Sevens, something better was coming.

In the meantime, (only six months after launch) both basic engine options were replaced by the then-new 997cc Anglia unit. This made a lot of sense, as Chapman's ties to Ford were growing – the Seven would have an all-Ford engine line-up, and the Lotus Cortina was on the horizon. Just as important, the new engine was revvy, tuneable and relatively cheap. So the new entry-level Seven was cheap too, at £499 in 1961 kit form; though you could have it without that engine for £100 less. It still cost the same two years later, when a mass-produced Sprite or MG Midget would set you back £587 and £599 respectively. The new Triumph Spitfire cost even more. Lotus must have been glad of its good value Sevens, which continued to sell well. The Elite, by contrast, cost only £167 less than an E-type. Until it was offered in kit form, unsold

**Above and right**

*Lotus Seven Series 2 built in 1962, chassis no. SB1204. 1500 Ford Cosworth engine. Ron Welsh bought this Series 2 in 1962 for £499 in kit form and built it over a long weekend – driving up to Lotus at Cheshunt for his free check. In 1965 he replaced the 1500 Cosworth for a 977cc Anglia engine to race in the up-to-1000cc classes and in 1970, after racing it successfully, he returned the 1500 engine and road equipment and still uses it today over 30 years later. Elan S2 wheels and hub caps are fitted, as are earlier alloy cycle wings from a Series 1 Seven – a relic from Ron's racing days*

Elites began to line up outside the factory. The popularity of the Seven, however, seemed assured.

If a cheap, basic sports car was part of the Seven philosophy, so was serious performance. In July 1961, the Super Seven returned, this time with a Ford engine. It was the 109E, a 1340cc relative of the one litre, but ready tuned by Cosworth. Two 40 DCOE Webers, a four-branch exhaust system and some cylinder head improvements helped it turn out 85bhp. That legendary road tester, John Bolster, tested a Cosworth Super Seven for *Autosport* in late 1961. He liked the engine's 'immense punch in the middle ranges', the 'phenomenally good' acceleration and 'phenomenally high' cornering powers.

Phenomenally variable would perhaps have better described the acceleration – Bolster achieved the 0-60 dash in 7.6 seconds, while a more sober *Motor* test of the same car recorded 8.5 seconds. *Sporting Motorist* claimed to have done it in 6.9, but *Road and Track* could only get 9.9 seconds out of a LHD Super Seven (which according to the spec sheet boasted an extra 5bhp over 8843 AR, the English test car). Confused? John Bolster wasn't – after a passing reference to all the usual Seven discomforts he concluded, 'If you mind about this sort of thing this car is too good for you, and you had better buy the dreary conveyance which you deserve. If you want shattering acceleration, the right sort of

**Above**
*Lotus Seven Series 2, 1962, with Cosworth Ford 1500 engine. Dashboard with rev counter well placed through period wood wheel*

**Above right**
*Lotus Seven 1500 Cosworth Ford S2 from 1962. Original interior; note speedometer in front of passenger seat*

**Right**
*Lotus Seven 1500 Cosworth Ford S2 with twin Weber carburettors. This is owned by Ron Welsh*

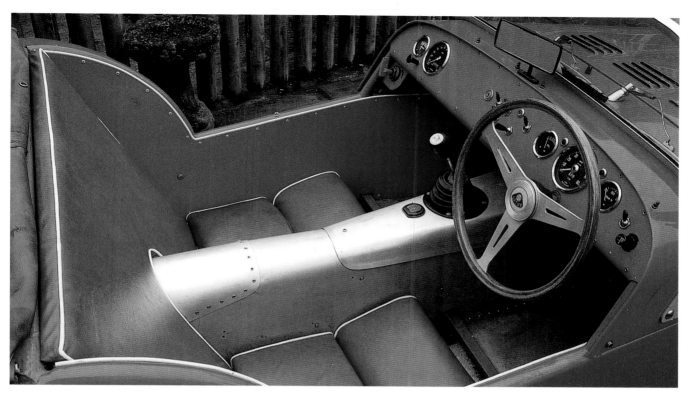

handling and a jolly good chance of passing the chequered flag while it is waving, the Lotus is your car.'

But the 1340cc Seven didn't last long. Only a year after John Bolster enjoyed his road test, the engine was superseded by the 1498cc five-main-bearing version straight out of the Cortina GT. You could have it in mild form (low compression, 66bhp, but lots of torque) as the Super Seven 1500, or of course, as a Cosworth. This time, all the usual modifications brought 95bhp at 6,000rpm. Just as well that both versions had front disc brakes as standard (for the first time on a Seven), while both also benefited from the Cortina GT's close ratio gearbox. But you still had to pay extra for flashing indicators and a heater, while Lotus appeared not to have grasped just how useful a fuel gauge was. The Cosworth 1500 accelerated faster than any other Seven, but it wouldn't do much more than 100mph – the aerodynamics were as bad as ever.

By this time, the Seven was getting decidedly middle-aged by Lotus standards. Five years after the original Series One, it was the longest-lived Lotus model of the time – the 6 had only lasted two years. Thoughts naturally turned to replacement, and they were to recur more than once in the coming years. In 1962/3 a project called 'M2' seemed to be the most promising. As ever, Colin Chapman wanted to cut costs and glass fibre unitary construction seemed the best way – no chassis or space frame to worry about, or any large pieces of body to join together. With suspension from a mass produced car, and the Ford Anglia engine, it should sell for the same price as a Midget, but out-perform it. Unfortunately, the 'unimould' idea didn't work – M2 gained a chassis, greater weight, a bigger engine...and became the Elan. For the time being, the Seven would have to soldier on.

*Lotus Seven Series 2 Ford 997cc engine, chassis no.1377. First registered on 20 February 1962 to an RAF pilot at RAF Wattisham, 266 Squadron. Bought by present owner John Green in August 1965 and used ever since in unrestored condition*

# Series Three
# Long-Lived Stopgap

Lotus Seven number three was never meant as more than an interim car – design work on the radically new S4 was already underway. Yet, courtesy of Caterham, it is the longest lived Seven of all, still in production in the 1990s, and thriving. With the S4 only a couple of years away, the Three was intended as a mild facelift, but there was another incentive to make changes.

Lotus Components, the subsidiary that made racers, was in trouble. It needed a profitable bread-and-butter car to even out the seasonal peaks and troughs of race car building. And with Lotus' mainstream road car business going ever upmarket, it made sense to separate the Seven. Mike Warner, a former Lotus man who had left to work for himself, came back to take charge. Perhaps his biggest achievement was persuading the Board to allow a small budget for Seven development.

They did well on that budget, for although the latest Seven (announced in September 1968) looked very like the old one, there were some major changes under the glassfibre and alloy. It didn't at first get the spaceframe strengthening which all S2's lacked, but that other notorious weak spot, the Standard Ten rear axle, was dropped in favour of a stronger, more modern item from the Ford Escort. This also brought bigger drum brakes (while discs became standard on the front). With a good choice of ratios, it was a welcome, if belated, move.

Ford by now had a monopoly on Seven engines, and the S3 offered a choice of 1297 or 1598cc crossflow engines from the Escort and Cortina respectively. The little one was an economy option, an attempt to keep the cheap Seven alive. In the event, of course, it didn't cost much less than the 1600, and most buyers chose the 84bhp Cortina engine. Despite no tuning, the 1500 proved just about as fast as the old Cosworth 1500, but was smoother and had a wider spread of power. Both engines typified the standard Seven units; relatively cheap and reliable, and you could buy spares anywhere. But if this wasn't enough, a

*Nearest is the Lotus 7 S3 1599cc Ford, built in 1969. Owner and driver Ian St. John accelerates past John Rees in his Lotus 7 Series 4 1600 GT at Mallory Park's Shaw's Corner hairpin*

Holbay-tuned version of the 1500 soon became available, giving enthusiasts a choice of three.

Other changes to the S3 were minor, but did represent the gradual encroachment of convenience engineering (something which has never overcome the Seven). The exhaust system became a more conventional set-up which exited at the rear; no longer would passengers risk scorched legs. The fuel tank filler, for so long part of the token luggage compartment went outside at last, while flashing indicators, seat belt mounts and an electric fan all became standard. 'Inside' (a doubtful term when applied to the Seven) there was a new dashboard featuring the longed-for fuel gauge, while the seats got a little more padding. To keep all this in perspective, the seats were still non-adjustable, and a heater was on the options list.

By 1969, Lotus' own Ford-based twin cam engine had been around for some time. Strong and powerful, the only real puzzle is why it took the company (now based at Hethel in Norfolk) so long to fit it to the Seven. But when it came, the Twin Cam SS was a real humdinger. With Holbay tuning, it gave a reliable 125bhp, which, although the Seven had put on a fair bit of weight since the S2 Cosworths was enough to give a blistering performance.

Perhaps even more significant than the engine were the chassis improvements that came with it. At long last, those tubes taken out of the S2's engine bay and side panels were restored. There was also

*Lotus Seven Series 3 Twin Cam 120bhp Holbay giving 230bhp per ton and 0–70 mph in 9.7 seconds, 0–60mph in 7.4 seconds and top speed of 112mph when tested by* Hot Car *magazine. The Holbay engine was balanced with special pistons, high lift camshafts, gas flowed head, new manifolds and twin 40 DCOE Weber carburettors*

strengthening for the bulkhead, transmission tunnel and for the steering rack mounts.

Road tests of this last of the S3s (it was introduced in October 1969, only six months before the S4 appeared) echoed those of every other Seven. *Autocar* found that it accelerated to 60mph faster than an E-type but the aerodynamics meant it barely cracked 100mph. *Road & Track*, with characteristic bluntness, described the SS as 'a 1200lb barn door.' But they also said that given the right conditions, 'the Seven will out-corner anything you'll ever meet.' Some things hadn't changed.

*Lotus Seven Series 3 Twin Cam SS Holbay Ford engine producing 120bhp giving 230bhp per ton. The Twin Cam SS is very rare as only 13 models were produced having their own chassis prefix letters SC before the numbers 2564 TC up to 2576 TC. This very collectable Twin Cam SS is owned by Vincent Haydon of Supersprint Cars, in Salisbury, Wiltshire who specialise in Lotus Seven restoration, parts and sales*

# Series 4
# Forwards or Backwards?

Replacing the Lotus Seven became a bit like redesigning the Mini. Various proposals came and went, but none succeeded. There was the M2, which became the Elan, then Colin Chapman decided that a cheap mid-engined car (his 'Council Estate GT40') was the way ahead. That turned out too expensive, and became the Europa. And there was the unimould all-glass fibre design, which eventually surfaced outside Lotus as the Imp-powered Clan Crusader. But by 1969 the new Lotus Components team under Mike Warner had made up its mind – the new Seven would be a much refined version of the old car; cheaper to make, more profitable and appealing to a wider audience. Just as important, they had the big chief's approval and blessing.

Given the go-ahead, the design team of Alan Barrett and Peter Lucas took just seven months to build a running prototype. A remarkable achievement, as it was virtually all-new. A cheaper chassis was the priority (according to one estimate, the S3's spaceframe cost twice as much to build as an Elan's backbone!) and the new car got a much simpler frame, relying more on welded steel panels than myriad tubes to give it rigidity. The earlier Seven 'body' was just a few panels and wings tacked onto the spaceframe, but the S4 had a proper glass fibre moulding which bolted onto the chassis. Not everyone liked the styling, but it was really a sensible update of the Seven theme, giving more space for occupants and luggage.

The suspension was all-new too, using Europa wishbones at the front, while at the back the old 'A' bracket (the cause of more than a few cracked differentials) was replaced by a four link system. The design team hoped to keep the now legendary handling, while providing a more comfortable ride as well. What didn't change were the engine options – Escort 1300 GT (68bhp), Cortina 1600 GT (84bhp) or Lotus twin cam (115 or 125bhp). Despite all the changes, the S4 weighed little more than its predecessor, though the price went up of course. The price of the new-for-1970 1600 was £895 in kit form, up £120 on the equivalent S3.

With the S4's modern styling, bigger body and (horror of horrors) sliding perspex window, for the side screens, there were inevitable cries from enthusiasts that the Seven had gone soft. The press liked it though;

**Above and opposite**
*Lotus Seven Series 4. Owner, Jonathan Leech of Bishop Stortford, bought and built the kit in 1972. This is the last Series 4 to leave the factory and beat the new VAT tax. Jonathan has covered 76,000 miles from new without any problems except a few oil leaks from the 1600 twin cam, big valve 126bhp engine. Chassis no. S4/3201*

Lotus Seven S4 1600 GT built in 1972, chassis no. S4/2746.1GT. Owner John Rees races the S4 in the Lotus 7 Clubs 'QED 7' Club Challenge Race Series. The S4's, despite being heavier than other 7's, are popular in racing as they are less collectable and therefore cheaper to buy

they enthused over the more ergonomic dashboard, and the fact that the new Weathershield hood kept out wind and water far more efficiently than the old one. Of course, the Seven hadn't become convenient – the seats were still unadjustable, and the Australian magazine *Motor Manual* pointed out that, 'you don't actually get into a Lotus Seven, you sort of put it on.' But it was just as fast as before, and the steering (2.8 turns lock to lock) still had more in common with a go-kart than a car. The magazine concluded 'Reborn 7 with stylish glassfibre body and better cockpit; tractable Cortina GT engine; exhilarating to drive; precise, agile handling; skittish in wet; excellent gearbox but heavy clutch; quite expensive as a start kit car'.

For a time it became fashionable to knock the S4. It was, said the critics, a car fallen between two stools – too soft for the Seven fanatics, too stark for the fun car brigade. And it wasn't cheap any more. Even in kit form it cost more than a Midget. But sales told a different story – according to most sources, around a thousand were built in just two years, a higher rate than for any previous Seven. And Lotus reputedly made a £150 profit on each one! But once again, Colin Chapman's upmarket aspirations had little room for the Seven, and in late 1972/early 1973 (no-one seems to agree precisely when) the axe finally fell. There would be no more Lotus-built Sevens.

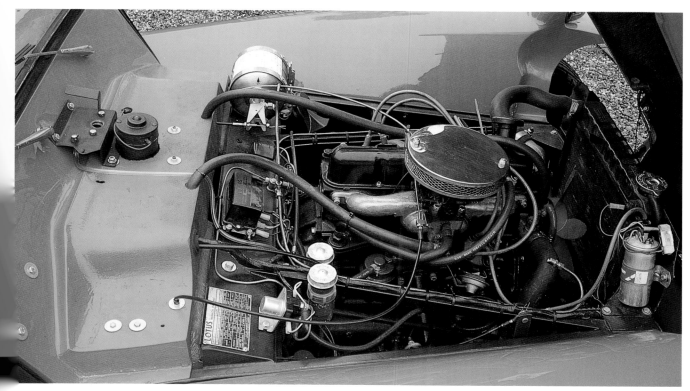

**Above**
*Lotus Seven Series 4 with 1700cc Ford 4-cylinder engine, built in 1972. Andy Shepherd racing his S4 at Mallory Park in the 'QED 7' Club Challenge*

**Right**
*The beautifully prepared 1691cc crossflow Ford engine in Wally Lile's Lotus Super 7 of 1962. The 4-cylinder Ford '711' block is very strong and reliable. Complete with four branch exhaust and twin Weber carburettors, maybe up to 180 bhp?*

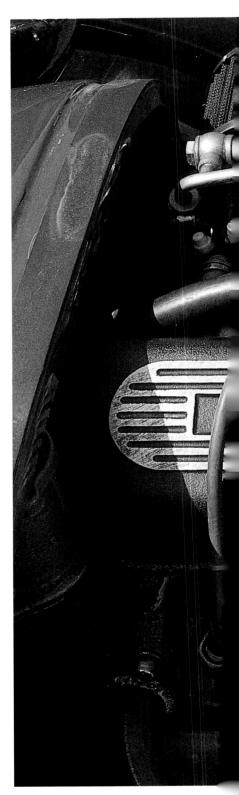

# Caterham
# Onward and Upward

Caterham Cars, based in a Kent village, had been one of the first 'Lotus Centres' but gradually came to specialise in Sevens. And from 1966, when founder Graham Nearn persuaded Chapman not to axe the little sportster, it had been the car's sole distributor. Graham Nearn lost no time when he heard that Lotus were ending production. In May 1973 the deal between himself and Lotus was made official. All spares, jigs, moulds and engines were sold to Caterham which would recommence production of the S4. After 14 years of servicing, selling and racing Sevens, Grahem Nearn knew the demand was there.

The enterprise almost stumbled at the first fence. After less than 100 S4s had been built, the hood supplier said it wanted £50,000 and a big order to keep supplying. It was out of the question, and production ceased. Fortunately, plenty of enthusiasts had been badgering Nearn to bring back the S3, and he obliged. Spaceframe assembly went back to Arch Motors (who had held the contract since the early S2 days) and the body reverted to its older, simpler style.

*'Arch': The Arch Company was started by Bob Robinson and Ted Young with the aim of building motorcycle racing sidecars in Tottenham, London in 1950. The workshops were situated underneath the railway arches – hence the name 'Arch'. The company built the tubular and box section chassis for the Series 1 Lotus Seven. Initially this series was built by Progress Chassis Company at Edmonton, London, but all production work eventually transferred to 'Arch'. Today they build the much improved Caterham '7' chassis and suspension parts. Over the years they have been instrumental in improving the rigidity of the Seven chassis which was not always possible with Colin Chapman's philosophy of light weight and cheapness. Arch also skin the bodies in aluminium to a very high standard – the same high quality that Williams & Pritchard achieved on the Series 1 Lotus Seven and Lotus 6. Today 'Arch', now called Arch Motor and Manufacturing Co. Ltd. of Huntingdon, Cambridge, have a wide base of work besides the building of Caterham chassis and body. They built the chassis for the first Lola Mk 1 Sports car, and now do many jobs for the successful Lola Racing Cars, also based in Huntingdon. They undertook the chassis engineering for the Ford RS200 Rally Car, and for most Formula Ford racing cars*

There were some worthwhile changes though, not least to the spaceframe itself, which standardised the strengthened version seen on the S3 SS. Among other things, Caterham also added reinforcement around the gearbox mounting, replaced the Escort rear axle with a beefier RS2000 one, and improved the seat belt mountings. And of course the car couldn't wear Lotus badges any more.

Engine options changed too – the 1600GT kept going, but its little brother was dropped and the twin cam (still supplied by Lotus) was standardised in big value 126bhp form. Engine supplies were to be something of a problem for Caterham, at least at first. Lotus stopped making the twin cam in 1975 when the Europa died. Then Vegantune built engines from Lotus kits, and when the supply of crankshafts dried up, Vegantune devised a 1598cc version using the later 225E block. The last straw, however, was Lotus running out of cylinder heads – too much even for Vegantune's ingenuity. Apart from the final 130bhp VTA version of 1979-80, the twin cam Seven was no more. It went out with a flourish though, and the Big Valve (that really was part of its official title) was faster than any previous production Seven.

Something else that changed was quality. The early Lotus kits' quality was a little variable. *Motor Sport* built a Cosworth Super Seven in 1963, and recorded a few teething troubles. The steering rack mounting holes had been drilled wrongly, the gear lever was too long and the carburettor and exhaust manifold flanges needed filing and cutting to fit. That was just the beginning though – the horn was vibrated to

*Chassis being welded together in the factory chassis shop. In a 1958 article for Motor Racing Magazine, Colin Chapman wrote: "Instead of reducing weight, it is better to design light in the first place, modifying any components which may show stress during development work. This, in the case of the frame, means having links which are all of the same strength."*

destruction, first one water pump failed, then another, then another, until a new electric fan eased the strain on its bearings. But not before the old fan lost one of its blades on the M1. Meanwhile, with the car run in and full power being used, the differential mounting bolts kept on working loose until they were drilled and wired. *Motor Sport* were actually quite fortunate with their kit – it wasn't uncommon to find some vital bit missing. And it was no myth that Colin Chapman wrote off the Elan's warranty costs as development spending.

Things went rather more easily for Jeremy Coulter, who wrote about building a 1600 Sprint in 1982 for *Thoroughbred & Classic Cars*. Apart from a minor exhaust manifold problem, his Seven went together easily and was on the road only five days after taking delivery. Of course, Caterhams are less of a kit than the old Sevens ever were – engine, gearbox and suspension are already in place and just need tightening up.

For one thing, the tax advantage has gone, with VAT payable on kits as well. But being sold part-assembled did exempt the car from expensive Type Approval regulations.

The Sprint engine in Coulter's car was really the successor to the old twin cam. With the Lotus unit unavailable, Caterham took the obvious step and started to offer tuned versions of the faithful pushrod 'Kent' engine instead. A peakier camshaft, high compression and the almost obligatory twin Webers produced 110bhp and 97lb/ft. Offering a lot more performance than the basic 1600GT at little more cost, it soon became the most popular option.

By 1985, it had grown to 1691cc and 135 bhp, which gave a 0-60 time of 5.6 seconds when *Autocar* tested a 1700 Super Sprint. For the very brave, Caterham now offered a 150 bhp BDR engine (Cosworth twin cam head on the Kent block). Still, even the Sprint accelerated faster

*Chassis parts being fitted in a factory jig. A fully detailed step-by-step guide to the assembly of the Seven can be found in* Lotus Seven *by Tony Weale (Osprey Automotive, 1991)*

**Right**
*Caterham chassis being skinned to give protection against corrosion*

**Below**
*Arch. Bonnet aperture for carburettor air filters being edged*

than anything else of the price. Not that Sevens in general were cheap any more – Auto Car's Sprint was listed at £7,769, without extras. On the other hand, that still made it cheaper than fully assembled rivals like the Morgan 4/4 or Panther Kallista. Like countless Seven road tests before, the Autocar once declared the car to offer more performance per pound than any other.

Fortunately, Graham Nearn and Caterham were not content to do a Morgan and go on producing the old favourite virtually unchanged. Throughout the eighties there were a number of significant changes under the skin (the one thing that would never be altered). Two inches extra leg room were liberated by moving the seat back – this was the Long Cockpit version – and an individually adjustable seat, joined the options list. A five-speed gearbox from the Ford Sierra replaced the Corsair 2000E one, and most significant of all, there was a completely new rear suspension from 1989. The de Dion system was independent, allowing a far better ride than the old live axle.

It was a heartening move, proving that Caterham knew what the car's shortcomings were, and were willing to make major changes to keep it up to date. It did come more expensive of course, but those buying a kit could still specify a live axle to keep the cost down. Then there was the HPC, a 15bhp projectile powered by a Vauxhall Astra 16v engine. Like the BDR, Caterham included a high performance driving course as part of the package. It brought far more publicity than any of the performance figures did. At the other end of the scale, Rover launched its up to the

minute K-series engine which was another new option. With standard catalyst, its chief bonus was exportability – some markets just wouldn't accept the Kent based engines any more.

In 1993, the 20th anniversary of Caterham's Seven takeover, the range was bigger than ever. Incredibly, among the ever more stringent noise and emissions regulations, the faithful pushrod Kent engine provides the mainstay. Available in four states of tune (84, 100, 110 or 135bhp) it is still the cheapest route to a new Seven. 'Cheap' is a relative term though; the 1600 Sprint cost £12,995 in kit form, and the 1700 Super Sprint £13,895. Caterham recognised this, and introduced the Super Seven Classic. With an 84bhp Kent engine, and reconditioned (rather than new) gearbox and rear axle, it sought to make a complete kit available at a much lower price. For those who couldn't afford £8,000 for one of these, there was still the option of buying a basic Starter Kit for just under £4,700, and searching scrap yards for the rest of the bits or buying them from Caterham when funds allowed. To its great credit, Caterham was trying to keep the Seven relatively affordable.

On the other hand, if you had money, a 1400cc K-series Caterham,

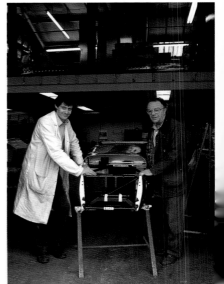

with De Dion rear end and 5-speed gearbox, would let you back £13,295 in kit form, or just under £15,000 fully built. The Astra 16v powered HPC was almost £20,000, ready for the road. But if its 165bhp weren't enough, 1992 saw the 'Jonathan Palmer Evolution' join the price list. With 250bhp from a tuned HPG engine, 1,988cc twin cam 16v, it was the most powerful Seven ever. And realistically, the most powerful there is ever likely to be. Yours for £33,750.

In the early 1990s, Caterham was secure. Just as it did 30 years before, the Seven was entertaining everyone who drove it. And with exports accounting for over half of what Caterham could produce (around 600 cars a year) it no longer depended on the vagaries of the UK market. The company even won a Queen's Award for Export in 1992, and Japan was one of the biggest customers.

**Above**
A Lotus Seven Club Brands Hatch Caterham 7 Super Sprint Ford engine in the Brands
Hatch pits, 1992

**Right**
Caterham 7 Super Sprint 1990 with the common modification by owners – the Lotus
Badge. Caterhams line up at the Brands Hatch pits for a drive around the circuit

**Left**

*Caterham Super Sprint at Brands Hatch circuit. In 1994, the HSCC QED Standard Road Sports Championship & Lotus Seven Club Challenge took place over eight days at Mallory Park, Brands Hatch, Lydden Hill, Cadwell Park, Croix en Ternois, Thruxton, Donington Park and Silverstone*

**Above**

*Caterham 7 Super Sprint at Brands Hatch. Note the metal strip on fibreglass nosecone to cut the timing gear for sprinting*

The Lotus Seven Owners' Club of Great Britain is split into regional areas, each organising its own meetings and events. There are regular gatherings for the club as a whole too, such as this day out at the Brands Hatch circuit in Kent. The club is part of a worldwide organisation for owners and enthusiasts of Lotus and Caterham Super Sevens. 95% of the cars here are Caterhams, the rest are Series 4 Lotus Sevens

**Above**
Caterham Super Sprint 1990 at Brands Hatch. For most Seven owners, the lure of the track is irresistible. The car is very user-friendly and quickly inspires driver confidence

**Right**
Caterham Super 7 with pretty cycle wings. 1990 HPC 7

**Left**
A Caterham Super Seven 1700 Super Sprint begins to take shape. Since Caterham's takeover in 1973, most Sevens have been fitted with Lotus twin cam engines and the majority have been exported

**Above**
Caterham factory at Dartford, Kent. Super Seven HPC, Vauxhall 1998cc four cylinder engine. The 16-valve twin camshaft engine gives 175bhp at 6000rpm

*Production line of Super Sevens at Caterham. The factory can produce about 600 units a year, of which over half are exported*

*Production line build up of Super Seven HPC Vauxhall 2 litre 16-valve. On completion, each vehicle is rigorously checked before its road test. Caterham recommends that all body-to-chassis mounting bolts are checked once a year*

A Caterham Super Seven 1700 Super Sprint 1691cc four cylinder engine which provides 135bhp at 6000rpm

**Left**
*Caterham Super Seven front suspension and disc brake. Note the top wishbone with front antiroll bar mounted to it in parallel. This is a big improvement on the Lotus Series 3 which had the single top link and antiroll bar*

**Right and following pages**
*Caterham Super Seven K-Series. The lightweight aluminium Rover 1.4 litre four cylinder engine has to be revved to produce its power but it does make this Caterham the best balanced Seven so far. The low weight of the Rover engine removes the characteristic Seven understeer which can be unpleasant in some Sevens despite the easy cancellation by power oversteer. The K-series Seven has light, precise steering and neutral handling with great grip which makes for safe, fast cross-country journeys. It is surprising that Colin Chapman placed the engine so far forward in the original design*

**Right**

*Caterham Super Seven K-series badge proclaims 16 valves. A powerful machine,
although perhaps its up to the minute K-series engine from Rover is better suited to a
more leisurely driving experience than the high performance Vauxhall Astra 16v*

*Caterham Super Seven K-series. Leather interior is an optional extra. The seats are comfortable and the driving position is superb. A far cry from the early days, when ownership of a Lotus was thought to be for only the hardiest of souls, as this extract from a Summer 1968 issue of the AA's magazine Drive demonstrates: "The Lotus Super Seven is very fast, very noisy, very draughty and very uncomfortable. The price of racing car performance is racing car exposure to the elements, a deafening side-mounted exhaust and a very cramped cockpit - in which you sit on hard, fixed cushions while your legs vanish into the hot pit under the bonnet to grope for the unseen pedals. Because it is little more than a road going racer, the Lotus has board-like suspension which gives a very hard ride"!*

*Caterham Super Seven K-Series, which comes with Caterham camshaft covers on the Rover 1397cc engine. This aluminium, twin camshaft, fuel injected 16 valve engine gives 103bhp on unleaded fuel at 6000rpm and torque of 127 Nm at 5000rpm. Fuel consumption averages out at 35mpg on mixed driving but 40.9mpg is the government figure at 75mph*

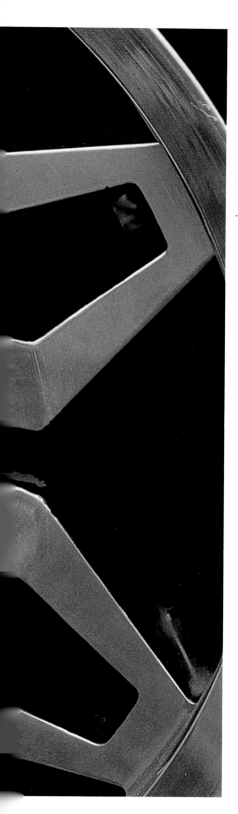

**Above**

*Caterham Super Seven K-Series, chassis plate Rover 1397cc engine but joined to a
Ford Sierra XR4i gearbox used on earlier 5-speed cars*

**Left**

*Caterham Super Seven K-Series 6JX14 aluminium KN alloy wheels with
185/60 HR14 tyres. In his 1958 Motor Racing Magazine article, Colin Chapman
wrote: " The total weight of a racing car on the starting grid, with driver and fuel
aboard, is in the limit determined by the weight of the engine; if this can be decreased
by one pound then the rest of the car can be made four pounds lighter. In all this,
however, tyres and wheels are the critical factor and their weight has a considerable
influence on the car as a whole. The job of the designer is to keep the right
relationship between sprung and unsprung weight and, above all, to keep the greatest
possible amount of tread in contact with the road."*

*Caterham Super Seven K- Series Rover engine has a power to weight ratio of 200 bhp per ton. Top speed is over 110mph with 0–60mph in less than six seconds*

*Caterham Cars' Anniversary Model
Super 7 sprint of 1992 at the
showrooms on the site of the old factory
at Caterham Hill*

# Some Caterham Rivals: Westfield Diesel

It is nice to think that Colin Chapman would have approved of a diesel powered Seven – not that he was too concerned about fuel economy, but he did appreciate efficiency; and a modern turbo diesel would certainly provide that. There never was a Lotus Seven TD of course, but something quite similar did appear from a thriving sportscar firm somewhere north of Birmingham. Chris Smith had started making Westfields, replicas of Lotus Elevens, in 1982. The business thrived, and within two years his second project was ready for the road – a true-to-form kit replica of the Seven. This too did well, and like its inspiration, the Westfield progressed to flared front wings and ever more powerful engines, including a 280bhp Rover V8. With its spaceframe and all-independent suspension, the only big difference from a later Caterham was the body – all-glassfibre rather than alloy panels.

Meanwhile, Richard Wilsher, owner of Sword Automotive in the tiny village of Mark on the Somerset Levels, was getting interested in diesels. Richard's background was in motor racing, having worked for some of the world famous teams, and he began to realise the tuning potential of oil burners after turbocharging a customer's Citroën Visa. The result was CIRCE – Compression Ignition Race and Competition Engine. The basis was one of the most common small diesels on British roads, the Ford 1.8 litre. It wasn't renowned as the most powerful or refined engine on the market, but the results obtained by Sword were remarkable.

A much larger Garrett T3 turbocharger and an air-to-air interercooler were fitted, both of them from the Sierra Cosworth. Maximum boost was raised and there were various internal modifications, such as reshaped pre-combustion chambers and piston crowns. Power shot up 40 per cent to 110bhp, and (the strong point of

**Right and overleaf**
*Westfield SE Diesel being tested by the editor of* Diesel Car *magazine John Kerswill. This is the only convertible diesel sports car. The driving experience is the same as in any Seven and includes the familiar bumps, draughts and leaks!*

any diesel) there was 150lb/ft of torque. CIRCE was a natural candidate for the Westfield – not only would its light weight make the most of the performance–economy balance, but a bright red diesel sportscar would create tremendous publicity. And it did. The Westfield Diesel was tested by just about every UK motoring magazine in 1992. *Autocar* claimed 0-60mph in 6.6 seconds and 30-70mph in 7.3, which would have made it about as fast as a twin cam Seven! It was also as frustrating as the Seven can be. The author was involved in testing the Westfleld for *Diesel Car* – a non-functioning speedometer and mileometer meant we could test neither acceleration nor fuel consumption. The hood then blew off in the middle of a maximum speed run. Based on estimated mileage, fuel consumption was around the mid-40s, well below the 55mpg Westfield thought was possible with mixed driving. Though it did provide the same driving experience as any Seven – the view down that long bonnet, between the big headlamp shells; the robust bellow from the silencer just below your elbow; and of course the bumps, draughts...

Westfield hoped to build replicas for around £12,500 fully built, but it wasn't to be. Cost was the main problem – Richard Wilsher estimated that CIRCE would cost around twice as much as an equivalent petrol engine to produce. That wasn't the end of the story. At the time of writing, Sword were in the process of extracting 130 bhp out of CIRCE to further demonstrate the power potential of diesel.

*Westfield SE Diesel 1800 Ford Turbo diesel engine*

**Above**

*Westfield SE Diesel 1800 Ford Turbo diesel engine with Garrett T6 Turbo-charger and intercooler. This engine produces 110bhp with massive torque of 150 lbs/ft.. This helps towards Westfield's aim of 30–70mph in 7.3 seconds. Richard Wilsher of Sword was responsible for tuning the Ford Diesel*

**Right**

*Westfield Diesel with characteristic Westfield leather steering wheel and dashboard*

**Above**

*Westfield Seight pretty alloy wheels with massive Goodyear 225/50 ZR15 tyres;*
*(see overleaf)*

**Right**

*Westfield Diesel prototype has smart all-red interior to match the exterior*

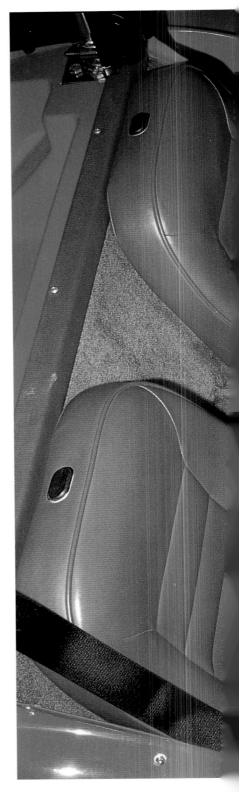

# Westfield Seight

The V8 engined Westfield Seight is a pocket-sized supercar. The 3.9 litre all aluminium Rover/Buick V8 engine of 285bhp in the lightweight Westfield, which weighs not much over half a ton, produces a power–weight ratio of over 400bhp per ton. There are not too many road cars around to compete with that. The acceleration from a standing start to 100mph is under nine seconds, which leaves the Porsche 911 Turbo, Lotus Esprit Turbo and Lamborghini Countach driver wondering what has happened to his supercar. There is one road going supercar that can out-accelerate the Seight – the F40 Ferrari, but that does cost ten times as much to buy and insure. Top speed of the Seight is about 140mph, which isn't bad for such an unaerodynamic car.

The in-gear acceleration times of the Seight are phenomenal with 50–70mph taking 2.5 seconds. That acceleration can deal safely with other traffic and leave the Westfield driver with a permanent grin. Great pleasure also comes from the sound of the exhausts mounted down the sides of the body – they produce a beautiful V8 note without being so loud as to be anti-social.

With the massive V8 performance and racing looks you would expect the Seight to be a disruptive, unacceptable sports car. However, it is not; it has a sensible chassis with balanced handling and good roadholding.

The first few hundred miles of driving will give the most arrogant of drivers or motoring journalists a chastening introduction. Once the initial alarm of having so much power is overcome you can exploit the Seight in complete safety. It is a very stable sports car but can be made to spin its rear wheels in fourth gear and even fifth gear in the wet. Given maximum acceleration on a straight road, the Westfield Seight stays straight and does not veer; on cornering more discretion is needed. However if adhesion is lost at the rear, the light, precise, quick steering can send the car back in the right direction.

The box section spaceframed chassis is light and rigid. At the rear, the well-located Ford Sierra Cosworth limit slip differential helps the traction of the wide 225/50 profile tyres. The Westfield independent suspension all round works well, though the ride is kept quite firm. The disc brakes all round give amazing stopping power on the large section

*Westfield Seight at speed. With over 400bhp per ton, wheel spin in third gear in the dry is possible*

tyres, with 100mph to zero in four seconds. Braking in the dry, the Seight is stable and well balanced. The aluminium V8 Rover/Buick engine weighs the same as a Ford Pinto 4-cylinder.

The Seight is a small car - only 12 feet long - but it can take six foot plus people in reasonable comfort. With the sidescreens up there is very little backdraught at the legal limit on motorways. You do not feel quite as vulnerable as might be expected because the all-round visibility is so good with the hood down.

The prototype Seight in the photos has a J.E Motors of Coventry tuned Rover V8 engine. They prepare the Land Rover and Range Rover racing engines for the Paris–Dakar race. The unit in the Westfield produces 285 lbs of torque at 4500rpm but even with the 4 Dellorto carburettors it will pull smoothly from 1000rpm, making city driving easy.

The standard production Westfield Seight has the fuel injection clean emission 200bhp V8 Rover engine and Rover 5-speed gearbox. This engine gives good power and better economy than the prototype engine.

The cost for the Seight kit with all new parts is £17,000 including taxes; the kit comes partially completed and should only take a day to be finished. Last year over £200,000 was spent passing strict EEC Approval including crash and seat belt anchorage tests.

Chris Smith of Westfield has been building sports cars for a decade. Today the company operates from a large modern factory in Dudley, West Midlands.

*Westfield Seight design is a derivation of the sports cars of the late 1950s, similar to the Lotus Seven. It has a modern running gear 3.9 litre Rover/Buick V8, five speed Rover gearbox and a Sierra Cosworth limit slip differential*

Despite its racing appearance, the Westfield Seight is a stable sports car with good handling and roadholding. Its acceleration is outstanding, with 50-70mph taking just 2.5 seconds. Only the F40 Ferrari, which costs ten times as much, can better this

*The tidy leather interior of the well thrashed press car. At the time of writing, the cost of the car in kit form with all new parts is £17,000 ($26,150). It arrives partially completed and can be on the road in less than two days*

*Ginetta G2 built in 1959 with a Ford side valve 1172cc engine. Built in kit form by Wilkett Brothers, Agricultural and Constructional Engineers, in Woodbridge, Suffolk. The kit was designed to compete with the successful Lotus Six and Seven. Production started in February 1958 and finished with an order in 1991 for six special versions for Japan. The G2 price in 1958 was £156 ex works*

# Donkervoort S8AT

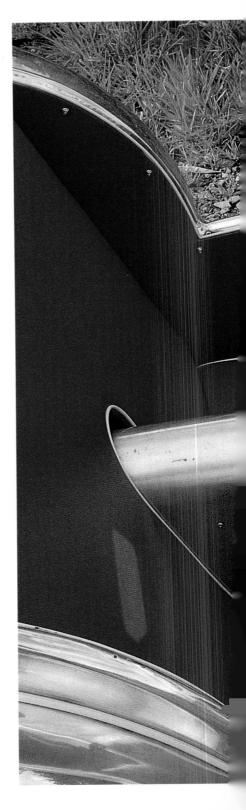

The ultimate classic shaped Lotus Seven-type sports car is built in Holland by Joop Donkervoort. He has been building and developing his cars for ten years in his small but spotless factory behind his house in the charming Dutch countryside at Loosdrecht near Utrecht.

Joop has built a very different machine to the original spartan and cheap classic Lotus Seven. His car has all the speed and roadholding but with the added bonus of a comfortable ride, rugged, reliable and user friendly. In the rain, the soft top fits and does not let water in if you are cruising on the Autobahn at 100mph. A modern car with a classic shape, the S8AT's space frame chassis carries Joop's own rear independent suspension and axle. The front suspension is similar to Caterham's, but with Donkervoort's own titanium and aluminium alloy uprights. Koni dampers and springs are used all round. The ride, handling and grip on all road surfaces is excellent. Joop's attention to detail and build quality is first rate. Nice details like the mounting of the battery at the rear of the car, away from the engine heat, also low down to keep the centre of gravity as low as possible; likewise the low mounting of twin aluminium fuel tanks. The total fuel capacity of over 45 litres gives a range of nearly 250 miles. Fuel consumption is very dependent on the driver, varying from 20 to over 30mpg.

A brake balancer control bar allows owners to adjust front and rear braking to their own choice.

The S8AT uses the single overhead camshaft four cylinder Ford 2-litre engine with a Garret T3 turbo and intercooler. This engine is cheap to service and very reliable, with lots of power and torque. The Ford electronic fuel injected engine also comes with three-watt catalytic converter which removes some of the poisons from the exhaust in return for a smell of bad eggs. This cat engine produces 170bhp, enough for 0–60mph in 4.8 seconds to put the lightweight S8AT in the supercar class. Top speed is not the car's best attribute, but 130mph is claimed by Joop for the S8AT.

A pleasant Ford 5 speed gearbox helps the acceleration times and the overdrive 5th gear makes open road cruising relaxed. The steering is nicely weighted and acceptably responsive but is not as good as the K-series Caterham. The wide tyres that help the roadholding deaden the steering. The 205/50 x 15-inch front and 225/50 x 15-inch rear tyres are mounted on pretty alloy Revolution competition wheels.

On the road the S8AT feels incredibly rigid with its substantial space frame chassis, which has deformable structures at the front and rear in case of accidents. The pretty aluminium panels are glued and riveted to the chassis which means that no rattles or shakes can be induced on bumpy roads. It is not everyone's idea of a fun sports car but the limited production S8AT sells well in Europe where the £25,000 ($38,500) price tag does not seem to deter enthusiasts.

# Useful addresses and phone numbers

Historic Lotus Register,
Membership Secretary,
Mary Hobson,
Badgers Farm,
Short Green,
Winfarthing,
Norfolk
IP22 2EE

Membership Secretary,
Lotus Seven Owners' Club,
BM Box 8248
London
WC1N 3XX

Mike Brotherwood,
Lotus 7 and Historic Lotus
Restoration.
0249 817338

Caterham Cars,
Seven House,
Townend,
Caterham Hill,
Surrey.
CR3 5UG
0883 346666

Chassis Builders,
Arch Motor & Manufacturing Co.,
Red Wings Way,
Huntingdon.
PE18 7HD
0480 459661

Derek Moore - Parts and
Restoration and Service,
The Classic Carriage Co.,
9 Somerfield Way,
The Pastures,
Leicester. LE3 3LX
0553 862499

Vincent Haydon,
Restoration Spares and Sales,
Supersprint Cars,
Harnham Garage,
Newbridge Road,
Salisbury,
Wilts.
0722 410077

Fred & Lee Fairman,
Lotus 6 & 7 chassis & body
construction,
Port Isaac Road Station,
Trelill,
Bodmin,
Cornwall.
PL30 3HZ
0208 850837

Donkervoort Automobiel
Produktie BV,
Nieuw Loosdrechtsedijk 205A
1231 KT Niew Loosdrecht,
Holland

Westfield Sports Cars Ltd,
1 Gibbons Industrial Park,
Kingswinford,
W. Midlands.
DY6 8XF
0384 400077

# Specifications: First and Last

| | 1957 Lotus Seven | 1992 Caterham Seven JPE |
|---|---|---|
| Price | £526 (kit form) | £33,750 |
| Engine | 1,172cc sidevalve | 1,998cc twin cam 16v |
| Power | 40bhp @ 4,500rpm | 250 bhp @ 7,750rpm |
| Torque | 58 lb/ft @ 2,600rpm | 186 lb/ft @ 5,250rpm |
| Kerb weight | 1,009 lb | 1,193 lb |
| Top speed | 81 mph | Faster |